Data, Chance & Probability

Grades 1-3
Activity Book

by Graham A. Jones
& Carol A. Thornton

⚠ WARNING:
CHOKING HAZARD - Small parts.
Not for children under 3 years.

Table of Contents

Introduction

This *Data, Chance, and Probability Activity Book, 1-3* contains reproducible blackline masters, teaching notes, and activity pages designed to help children in grades 1-3 explore mathematical ideas about outcomes, likelihood, data and its presentation, and probability. The activities have been carefully developed to engage children in exploring their own real world, especially the children's school environment.

The focus of activities is on enhancing problem-solving and communication skills. Individual activity pages can be used by children working in pairs, with small groups, or as an entire class. Activities are slightly more difficult as children progress through a section; the increasing level of difficulty is more apparent from section to section. Teaching Notes at the beginning of each section offer easy-to-follow guidelines for every activity page, from materials needed to assessment of student progress.

There are 54 one-page activities, presented in eight sections. In the beginning of each section are Teaching Notes covering three aspects of using the activities:

Warm-Up
Prerequisite activities that set the stage for the activity pages.

Using the Pages
Comments on each activity page, dealing with the purpose and possibilities of that page.

Wrap-Up
Activities to check for understanding or use of new terms.

How to Use Activity Pages

Each student page presents a hands-on activity. In the first part of an activity, students address a problem, or "Explore." They have clear, brief instructions and use grab bags, spinners, and other simple manipulative materials. In the second part of the activity, children communicate their results by discussion or in writing. Their instructions are either to "Talk About It" or to "Write What Happens." In the third part, students interpret and extend the ideas in a "Think and Tell" exercise, reflecting on what they've done and applying their ideas and results.

The three parts of each activity build problem-solving skills. First, children have an opportunity to address a problem and predict results. Then they experiment, collect and discuss data, and share their ideas and results in small groups. Finally, in a larger class group, they can discuss their conclusions, their interpretations, and the applications of their ideas. The three parts of each activity page are clearly labeled with words and icons, as shown.

1. Address a problem.

3. Interpret and extend ideas.

2. Communicate results.

or

NCTM Standards

The foundations for many mathematical ideas are in this *Data, Chance, and Probability Activity Book, 1-3* including outcomes and events, likelihood and chance, data and its presentation, probability, and using math to express probabilities. These ideas are discussed only in the Teaching Notes for the activities. It is generally left to you to introduce or use terms such as *outcome, data,* and *likelihood* when helpful and appropriate.

The mathematical ideas can be expressed in terms of several skills and subskills:

• **Describe an outcome.** Children can describe simple events and outcomes, and distinguish between several possible outcomes. (Example: Children place bears of various colors in a bag, close their eyes, and select a bear. The outcome is the color of the selected bear.)

• **Describe data.** Children can describe several outcomes as a list, a set of tallies, or a graph. (Example: Children spin a spinner four times and tell the outcomes. Those data, or outcomes, can be summarized and presented as a tally or other list.)

A related skill is that children can look at a situation involving a spinner or tossing a coin, and tell whether that situation seems fair or unfair (that is, whether each outcome is equally likely or all outcomes should happen an equal number of times).

• **Use numbers.** Children can use tallies, graphs, or numbers to describe data. For example, children tally the "favorite bear" of each member in their class, and tell the number of votes for each bear. Using graphs to show the votes is called "data presentation"; talking about these graphs is one form of "data analysis"; using numbers to compare and analyze situations or data is the first step of a skill often called "mathematical modeling."

The probability data can then be used to explain why a situation is fair or unfair. Children can also use the data to make probability comparisons to tell whether one outcome is more or less likely than another.

Two recent Standards documents of the National Council of Teachers of Mathematics, including the *Standards on Probability and Statistics,* call for activities like the ones in *Data, Chance, and Probability Activity Book, 1-3*. The activities focus on problem solving, reasoning, communication, and making connections. This data and chance book can be a useful supplement to existing mathematics programs in regular classrooms and special education or remedial math settings.

Activity Masters and Game Boards

The last section of this book contains seven activity masters, which are used in about 15 of the activities. Each activity page contains a box listing the specific activity master and other needed materials.

The last section also contains a Progress Chart, which is a record-keeping form to help keep track of students' skills developed while using the activity pages. An Award Page, a Family-Gram, and an Award Certificate help motivate and encourage students' work.

Assessment

Assessment should be a natural part of instruction—before, during, and after an activity. This assessment should focus on children's thinking, problem solving and understanding of ideas, and reactions to learning about data and chance. Teachers should assess progress by observing and monitoring the children's actions, and their oral and written responses. The "Talk About It," "Write About It," and "Think and Tell" components of an activity provide special opportunities for assessing children's thinking about data and chance and the impact of the learning experience.

In fostering understanding and problem-solving skills for young children, teachers need to assess students' abilities to THINK and VERBALIZE as they:

- **Explore a problem.**
- **Predict the outcome.**
- **Model it in different ways.**
- **Gather data or information.**
- **Organize the information.**
- **Construct and interpret data display.**
- **Validate their conclusions.**
- **Create their own problems.**

These actions build a strong thinking and knowledge base for mathematics. Teachers also should assess whether children are able to correctly use the language of data and chance, especially when they work in small groups and whole class settings. Important vocabulary for data and chance includes the following:

New Terms

- Certain
- A chance/possible
- No chance/impossible
- Best/least chance
- Outcomes
- Equally likely outcomes
- More likely
- Most likely
- Least likely
- Fair die or spinner
- Data
- Graph
- Bar graph
- Tally
- Most frequent score (mode)
- Smallest-biggest (range)
- Middle score (median)

Children will not learn at the same rate, nor will they investigate activities to the same depth. In their enthusiasm for working with probability and chance, they will generate many questions that often will reveal the breadth as well as the depth of their thinking. Since assessment should guide further instruction, information gained from observing and monitoring children may lead to extension activities created by the teacher or even students.

Bears & Spinners
Teaching Notes

Materials Needed: Bags for Bears • Red and Blue Bears •Paper Clips for Spinners

Warm-Up

The six activities in this section introduce children to the ideas of outcomes and events, and emphasize that an outcome may be certain, possible, or impossible. The activities are designed to help children recognize that if a particular outcome always happens, it is certain; if it never can happen, it is impossible; and if it can sometimes happen, and sometimes not happen, it is possible.

A simple introduction would be to talk about events that:

• always happen at school (books are used; the teacher writes on the board; children sharpen their pencils);

• sometimes happen at school (the principal visits the class; there is a room party; the class goes on a field trip); or

• never happen at school (someone brings a lion to school; a spaceship lands in the playground; Christopher Columbus visits the school).

Children might be encouraged to draw or cut out pictures for a class bulletin board to illustrate certain events, possible events, and impossible events.

Using the Pages

All the activities in this section give children opportunities to describe an outcome. You may want to start using the term *outcome* after the first few activities.

Bears in a Bag *(page 9)*: Be sure that children do not peek inside the bag as they pick a bear. Encourage children to use the words *always*, *sometimes*, or *never*, and make the links with the ideas of certain, possible, and impossible events.

Pick a Bear and **Pick Again** *(pages 10 & 11)*: An important difference between these two activities is whether or not the children replace the bear in the sack after selecting a bear.

Pick One More Time *(page 12)*: Unlike the two previous activities, these two bears have different colors, so the outcome (the color selected) is not certain.

Spin and See *(page 13)*: Children may need help in learning to use the spinner. Make sure the spinner circle is kept flat on the table while the paper clip is being spun.

Spin and Tell *(page 14)*: The outcome "two" is certain, but it is not certain whether the outcome is the figure "2" or two objects.

Wrap-Up

To extend their ideas of certain, possible, or impossible events from the classroom to the home, ask children to describe and discuss events that always, sometimes, or never occur at home.

Bears in a Bag

Name_____

 **xplore** Take turns.

Close your 〜〜 〜〜 .

Pick a bear. Then open your eyes and tell the color.

Put the bear back.

 **Talk About It!** Tell what happens:
- sometimes
- always
- never

 **Think and Tell!**

Is it always certain to be a bear?

Need: Bag with 1 red and 1 blue bear.

Data, Chance, and Probability Activity Book, 1-3
© 1992 Learning Resources, Inc.

Pick a Bear

Name_____

Explore   Take turns.

Close your 〜〜 〜〜 .

Pick a bear. Then open your
eyes and tell the color.

Put the bear back.

Talk About It!   Tell what happens:
- sometimes
- always
- never

Think and Tell!

What color bears do you think are in the bag?

Check to see.

Tell what you find.

Need: Bag with 2 red bears.

 10

Data, Chance, and Probability Activity Book, 1-3
© 1992 Learning Resources, Inc.

Pick Again

Name_____

xplore Take turns.

Close your  .

Pick a bear. Then open your eyes and tell the color.

Don't put the bear back.

Tell what happens:
- always
- never

If you put the bear back and pick again, are you certain to pick a red bear?

Need: Bag with 2 red bears.

Data, Chance, and Probability Activity Book, 1-3
© 1992 Learning Resources, Inc.

Pick One More Time

Name_____

 **E**xplore Take turns.

Close your 〜〜 〜〜 .

Pick a bear. Then open your eyes and tell the color.

Don't put the bear back.

Talk About It! Tell what happens:
- always
- never

Think and Tell! If you put the bear back and start again, is red certain on the first pick?

Need: Bag with 1 red and 1 blue bear.

Data, Chance, and Probability Activity Book, 1-3
© 1992 Learning Resources, Inc.

Spin and See

Name_____

 **E**xplore Take turns.

Spin the spinner.

Tell what you get.

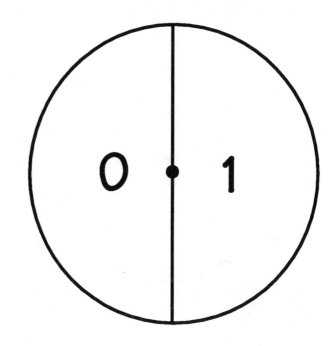

 Tell what happens:
- sometimes
- never

 Tell something about school that happens:
- sometimes
- never

Need: Paper clip for spinner.

Data, Chance, and Probability Activity Book, 1-3
© 1992 Learning Resources, Inc.

Spin and Tell

Name_____

 xplore   Take turns.

Spin the spinner.

Tell what you get.

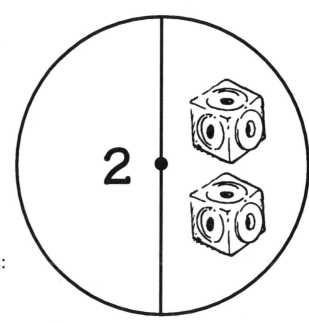

 Tell what happens:
- always
- sometimes
- never

 Tell something about school that happens:
- always
- sometimes
- never

Data, Chance, and Probability Activity Book, 1-3
© 1992 Learning Resources, Inc.

Hide & Spin
Teaching Notes

Materials Needed: Bags • Red and Green Bears • Red and Blue Crayons • Paper Clips for Spinners

Warm-Up

The four activities of this section move from certain events and impossible events to situations where children can make probability comparisons. For example, children are faced with situations where two events are each possible, but one is more likely than the other. Two of the four activities ask children to pick a spinner that would correspond to a real-world situation.

Because this section focuses on helping children recognize that some events are more likely to happen than others, a simple introduction might be to involve children in a "What's More Likely?" activity. In response to questions like the following, children stand to show their vote.

Sample Question 1: On a rainy school day, which is more likely?

Students will have umbrellas./Students will wear swimsuits.

Sample Question 2: On a cold day, which is more likely?

Students will order hot lunches./Students will bring their lunches.

The vote should be one-sided for Sample Question 1, because umbrellas are much more likely than swimsuits. The vote for Sample Question 2 may be closer because both outcomes are reasonable.

Using the Pages

Which Color? *(page 16)*: Children describe data (ten outcomes) by tallying. It may be necessary to review the idea of tallying with children.

Which Spinner Color? *(page 17)*: This is a "modeling" activity, because the spinner with 3-regions-to-1-region should give results similar to a bag with 3 red bears and 1 green bear. Care should be taken to note that the correct spinner has two outcomes — to parallel the two colors for the bears — and the ratio of the spinner colors is 3 to 1, just like the bears.

Which Color Now? and **Which Spinner?** *(pages 18 & 19)*: This pair of activities is similar to the previous pair. Children describe data (the colors of crayons), then they look at spinners that give the same ratio of outcomes.

Wrap-Up

In the last two activities, one event is twice as likely as the other. To help children explore and understand the notion of "twice as likely," use the data collected on page 18 to generate a class tally for red and blue. Then use cubes to make a bar graph, one cube for each tally, to show the total number for red and for blue. Children can observe that the bar for red is about twice as tall as the bar for blue.

Which Color?

Name_____

 **xplore** Put the bears in a bag.

Which has a better chance if you only pick one? Circle one:

red   green

Hide and Take Activity

Try picking a bear 10 times to see what color you get.

Put the bear back and shake the bag each time.

Tally what you get.

 RED _____    GREEN _____

Did your picks turn out as you said?

Why did one color have more tally marks than the other?

Need: Bag with 1 green and 3 red bears.

Data, Chance, and Probability Activity Book, 1-3
© 1992 Learning Resources, Inc.

Which Spinner Color?

Name_____

 **Explore** Which spinner works like the bears in the *Hide and Take Activity* (page 16)?

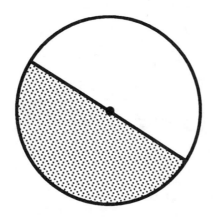 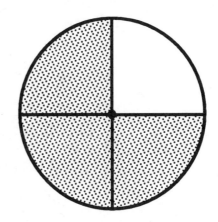

Talk About It!

Tell why.

Think and Tell!

Would this spinner also work like the bears in the *Hide and Take Activity* (page 16)?

Why?

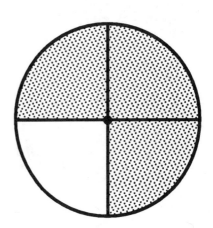

Need: Paper clip for spinners.

Data, Chance, and Probability Activity Book, 1-3
© 1992 Learning Resources, Inc.

Which Color Now?

Name_____

Explore Put the crayons in a bag.

Mix them up.

Which has a better chance if you pick only one? Circle one:

red   blue

Hide and Take Activity

Try picking a crayon 10 times to see which colors you get.

Put the crayon back each time.

Tally what you get.

 _____    _____

 Did the color turn out as you said?

 Why is red picked more often than blue?

Need: Bag with 1 blue and 2 red crayons.

Which Spinner?

Name _____

 **xplore** Which spinner works like the crayons in the *Hide and Take Activity* (page 18)?

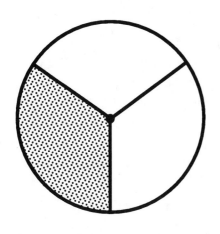

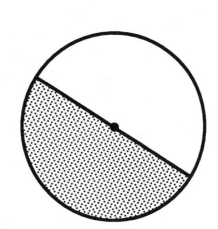

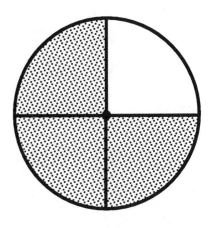

 Tell why.

 Would this spinner also work like the crayons in the *Hide and Take Activity* (page 18)?

Why?

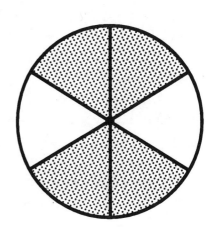

> Need: Paper clip for spinners.

Data, Chance, and Probability Activity Book, 1-3
© 1992 Learning Resources, Inc.

Bears, Books, Crayons, & Coins

Teaching Notes

Materials Needed:
Masking Tape • Square Pieces of Paper • Shelf of Books • Bags • Red, Green, and Blue Bears • Red, White, and Yellow Crayons • Red/White Counters (Chips) • Paper Clips • Nickels • Scissors • Activity Master 1

Warm-Up

For the first three activities in this section, children focus on collecting, presenting and using data to predict the likelihood of an event. Ask children to cut out graphs in old journals or newspapers, and display them on a class bulletin board. Discuss with children how graphs use numbers to present and describe information.

Also, children investigate the likelihood of events and use data as a basis for making probability statements. To introduce activities 3-4 through 3-11, ask children: "Which of the following is the most likely for travel to school?"

• Bus   • Car   • Walk   • Bicycle

(One way to answer the question is for children to make a tally or graph as a basis for determining which one is most likely.)

Using the Pages

Story Time (*page 22*): Three stories that have been read recently are needed for this activity. To prepare for the graph, the adhesive side of the tape should face out so the paper squares can be put in place quickly.

What Would the Story Be? (*page 23*): When trying to determine the story that is the most likely to be chosen by a student who missed the stories, children should refer to the data collected during the previous activity. The most likely story for the absent student is that which had the greatest vote. (Some children may want to know if the absent child is a girl or a boy.)

In the Library (*page 24*): It may be helpful to visit the library in advance and select appropriate rows of books. Alternately, rows of books on a classroom shelf might be used.

What's in the Bag? (*page 25*): In this activity children describe data, and in the next three activities, they compare these given numbers of colors with data they generate.

Best Chance and **Second Best** (*pages 26 & 27*): Children use numbers from the data in the activity, "What's in the Bag?" to predict the most likely color and the second most likely color. In the "Think and Tell" section, children might reflect on the total situation in different ways: blue is more likely than red; blue is more likely than green; green is more likely than red.

Which Spinner for the Bears? *(page 28)*: The spinner has 12 sectors because there are 12 bears. Children should color 2 sectors red, 4 sectors green, and 6 sectors blue. In the third part of the activity, children are asked to think about some of the different ways the spinner can be colored, as long as there are 2 red, 4 green, and 6 blue.

What Will It Be? and **Now What Will It Be?** *(pages 29 & 30)*: These two activities give children an opportunity to discuss "impossible outcomes." The color *yellow* is a choice for an outcome, but there is no yellow crayon and no yellow counter (chip).

Heads or Tails? and **Match the Coin** *(pages 31 & 32)*: In the activity "Heads or Tails," which uses the coin cut-outs from Activity Master 1, children generate data by tossing a coin, and then they display and describe their data. In the activity "Match the Coin," they look at spinner models for coin tossing.

 Wrap-Up

Given the discussion of the previous activities, children might be invited to write about their own 10 Tosses Graph in comparison to the Class Coin Toss Graph. Some children might note that the matching spinner in the activity "Match the Coin" is subject to the same extremes as the coin, but in the long run has the same probabilities as the coin.

Repeat the activity "What's in the Bag?" as a wrap-up activity because it combines predicting and analyzing data. Provide opportunity, such as a class bulletin board, for children to listen to or read others' write-ups.

Story Time

Name_____

 **E**xplore Think about the 3 [Stories] that were read.

Which did you like best?

Data Activity

Use your square to vote.

Put it on the tape strip for the story you picked.

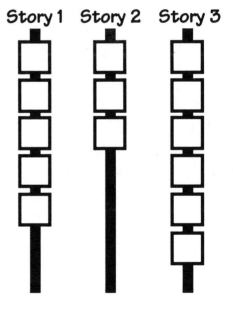

Story 1 Story 2 Story 3

 What does the graph tell?

 Did the girls vote the same way as the boys?

> Need: 3 stories previously read to/by class; 1 small paper square for each, white for girls, another color for boys; 3 tape strips sticky side up.

Data, Chance, and Probability Activity Book, 1-3
© 1992 Learning Resources, Inc.

What Would the Story Be?

Name_____

 If someone misses all 3 stories, what is he or she:

most likely to pick as the favorite?

      (pick one)

least likely to pick as the favorite?

      (pick one)

 Why do you think so?

 How does the graph from *Story Time* help you to decide?

> Need: Book and graph from *Story Time* (page 22).

In the Library

Name_____

 **Explore** Pick a row of 20 books.

Is there a book with more than one copy?

Is there an author who wrote more than one book?

 Why does the library need more than one
copy of some books?

 If you picked another shelf to explore, would you get
the same answer?

Need: Library shelf.

Data, Chance, and Probability Activity Book, 1-3
© 1992 Learning Resources, Inc.

What's in the Bag?

Name_____

Explore Look in the bag. About how many bears of each color are in the bag?

Color your estimate below.

Count the bears and color the graph to match.
Compare the graph with your estimate.

Red	my estimate							
	actual count							
Blue	my estimate							
	actual count							
Green	my estimate							
	actual count							

Think and Tell!

How do the colors in the graph compare?

Need: Bag of bears — 2 red, 4 green, 6 blue.

Data, Chance, and Probability Activity Book, 1-3
© 1992 Learning Resources, Inc.

Best Chance

Name_____

 **xplore**   Shake the bag.

If you pick one bear, which color is most likely?

Try picking a bear 10 times to see which colors you pick.

Put the bear back and shake the bag each time.

Tally what you get.

 RED _____    YELLOW _____    GREEN _____

Talk About It!   Did the color turn out as you said?

Think and Tell!   What does the graph say is the most likely color?

Need: Bag of bears and graph from *What's in the Bag?* (page 25).

 26

Data, Chance, and Probability Activity Book, 1-3
© 1992 Learning Resources, Inc.

Second Best

Name_____

 **E**xplore Shake the bag.

If you pick one bear, which color has the second best chance?

Try picking a bear 10 times to see which colors you pick.

Put the bear back and shake the bag each time.

Take turns and tally what you get.

 RED _____    YELLOW _____    GREEN _____

 **Talk About It!**

Did the second most likely color turn out as you said?

Did it turn out like the graph?

 **Think and Tell!**

Finish the story.

_____is more likely than_____ .

_____is more likely than_____ .

Need: Bag of bears and graph from
What's in the Bag? (page 25).

Which Spinner for the Bears?

Name_____

 **xplore**   Color the spinner so it works like the bears in the bag.

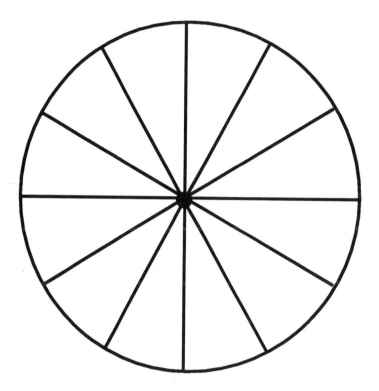

Talk About It!

Tell why your spinner works like the bears in the bag.

Think and Tell!

Would this spinner also work?

Need: Paper clip for spinner and *What's in the Bag?* graph.

Data, Chance, and Probability Activity Book, 1-3
© 1992 Learning Resources, Inc.

What Will It Be?

Name_____

 **Explore** Put the crayons in a bag.

If you take just one crayon, what color could it be?

Red	yes	no	maybe
Yellow	yes	no	maybe
White	yes	no	maybe

Crayon Activity

Try picking the crayons 10 times to see which colors you pick.

Put the crayons back after each pick.

Take turns and tally what you get.

 _____  _____  _____

 **Talk About It!**

Did the color turn out as you said?

 **Think and Tell!**

Does each of the three colors have a chance to be picked?

Need: 3 crayons-1 red, 1 white and 1 yellow.

Data, Chance, and Probability Activity Book, 1-3
© 1992 Learning Resources, Inc.

Now What Will It Be?

Name_____

 **Explore** If you toss the chip, what color could land face up?

Red   yes   no   maybe

White   yes   no   maybe

Yellow   yes   no   maybe

Chip Activity

Try tossing the chip 10 times to see which colors land face up.

Take turns and tally what you get.

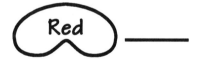 _____    _____

Talk About It!

Did every color have a chance?

Think and Tell!

How are the crayon colors from the last activity, and the chip colors in this activity:

- alike?

- different?

Need: 1 red/white chip.

Heads or Tails?

Name_____

 xplore Toss the nickel 10 times.

10 TOSSES GRAPH

										Total
Heads										
Tails										

Class Activity

Cut out or to match the graph.

Tape your coins to make a class pictograph.

How do heads compare
with tails?

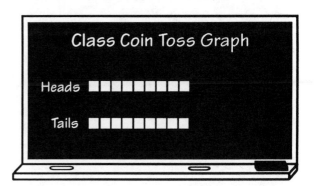

How are the class and 10 tosses graphs:

- alike?

- different?

Need: 1 nickel, scissors, tape,
Activity Master 1.

Match the Coin

Name_____

 xplore   Which spinner works like a coin?

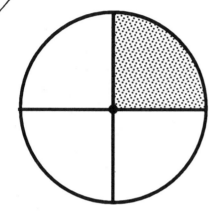

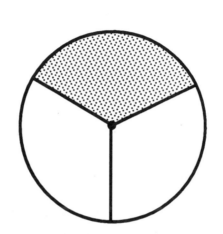

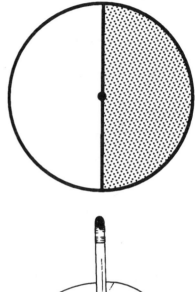

 Talk About It!

 Tell why.

 Think and Tell!

Would this spinner also work like a coin?

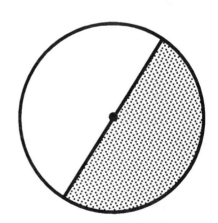

Data, Chance, and Probability Activity Book, 1-3
© 1992 Learning Resources, Inc.

Game Time
Teaching Notes

Materials Needed: Red/White Counters (Chips)• Dice (some with 2 red, 4 white dots; some with 1 red, 5 white dots; some with 3 red, 3 white dots) • Extra Red Dots • Paper Clips • Red and White 2" x 2" Squares • Masking Tape • Scissors • Activity Masters 3, 4, and 5

Warm-Up

In this section, children play games and discuss which game situations are fair. To introduce children to a mathematical meaning of "fairness," break the class into four groups. Make a rectangular game board. Draw in two diagonals to form four sections marked 1, 2, 3, and 4. Each group is assigned a section of the game board formed by masking tape on the floor.

An object (a checker, a bear) is dropped above the center point, and a group gets a point every time the object lands in their sector. The first team to get 10 points wins the game. Focus on the questions: "Is the game fair?" and "Does every group have the same chance on each toss?" As a follow-up, students might be invited to change the game board to make it fair.

Using the Pages

Is It Fair? *(page 34)*: Before students write about whether the game is fair, they should predict the results if the game is played several more times.

Make It Fair *(page 35)*: Be aware that there is still an element of chance operating in fair games. That is, one person could still win more than half the time in a fair game.

Which Spinner Color? *(page 36)*: In helping children to understand the modeling of a fair spinner, encourage them to find other ways of shading a fair spinner. Help them see that each spinner is divided into six regions. Then, if any three of the six regions are shaded, no matter what the order of shaded and unshaded regions, the total shaded area is 3/6 or 1/2 of the total area. So the spinner should act like a fair die.

Red or White? and **Try Again!** *(pages 37 & 38)*: If possible, retain the Class Graph in the activity "Red or White?" so children can contrast this chance situation with the one in the activity, "Try Again!"

Get a Match! *(page 39)*: In this game, a Wild Card will match any card. Therefore, a match is certain if a Wild Card is drawn.

Spin and Sum *(page 40)*: Children generate two numbers and describe possible, certain, and impossible sums.

Wrap-Up

Challenge children to change the Animal Snap Game in the activity "Get a Match!" so that the new game is fair.

Is It Fair?

Name_____

 xplore   Race Home Game

> Pick red or white. Put chips on start.
> Take turns. Roll the die.
> If red, move (red) one space.
> If white, move (white) one space.

Circle the winner: red   white

Does the game seem fair?

Make the Die Fair Activity

Change chip colors and play again.

Circle the winner: red   white

Does the game seem fair?

What makes a game fair?

Which color is more likely to win? Why?

> Need: 2 red/white chips; die with 2 red, 4 white dots; Activity Master 3.

Make It Fair

Name_____

 **xplore**   ## Race Home Game

> Pick red or white. Put chips on start.
> Take turns. Roll the die.
> If red, move one space.
> If white, move one space.

Circle the winner: red   white

Does the game seem fair?

Make the Die Fair Activity

Change the dots on your die to make it fair.

Play again.

Circle the winner: red   white

Talk About It!

Does the game seem fair?

What makes a game fair?

Think and Tell!

Can you lose a fair game? Why?

Need: 2 red/white chips; die with 2 red,
4 white dots; Activity Master 3.

Data, Chance, and Probability Activity Book, 1-3
© 1992 Learning Resources, Inc.

Which Spinner Color?

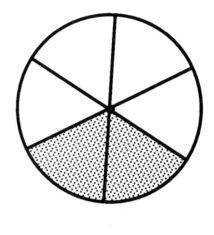

Name_____

 Explore Which spinner works like your fair die?

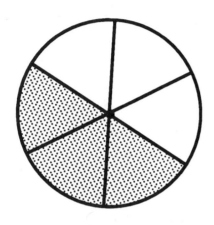 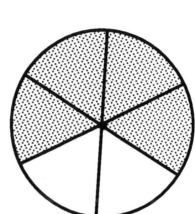 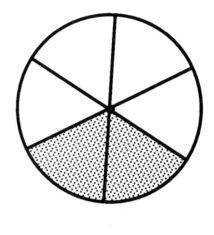

 Tell why.

 Is this a fair spinner?

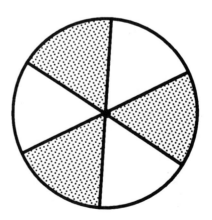

Need: Paper clip for spinners.

Data, Chance, and Probability Activity Book, 1-3
© 1992 Learning Resources, Inc.

Red or White?

Name_____

 **E**xplore

Race Home Game

Pick red or white. Put chips on start.
Take turns. Roll the die.

If red, move (red) one space.

If white, move (white) one space.

Circle the winner: red   white

Class Activity

Play at least 8 games.

Pick a square to match the winning color.

Tape up the winning color squares to make a class bar graph.

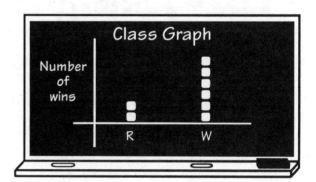

Class Graph

Number of wins

R   W

Talk About It!

How do red and white wins compare?

Think and Tell!

From the class graph, do you think the die is fair or unfair?

Need: Class supply of 2" x 2" red and white squares; tape; 2 red/white chips; unfair die (1 red, 5 white dots); Activity Master 3.

Try Again!

Name_____

 xplore

Race Home Game

Pick red or white. Put chips on start.
Take turns. Roll the die.

If red, move one space.

If white, move (white) one space.

Circle the winner: red   white

Class Activity

Play at least 8 games.

Pick a square to match the winning color.

Tape up the winning color squares to make a class bar graph.

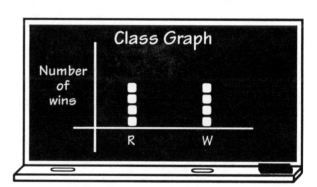

How do red and white wins compare?

From the class graph, do you think the die is fair or unfair?

Need: Class supply of 2" x 2" red and white squares; tape; 2 red/white chips, unfair die (1 red, 5 white dots); Activity Master 3.

Get a Match!

Name_____

 **xplore**   Cut out and sort the cards.

Make matches if you can.

Animal Snap Game

Take turns.
Mix all the cards and put them face down.
Pick one.
Partner tells if a match is:
* certain
* possible
* not possible
Turn the card over to check.

Talk About It!

When is a match certain?

Think and Tell!

What would happen if you played with this deck?

> Need: Activity Master 4.

Data, Chance, and Probability Activity Book, 1-3
© 1992 Learning Resources, Inc.

Spin and Sum

Name _____

Explore   Spin.

Partner spins.

Add your two numbers.

What could the sums be? Write them.

3

Talk About It!

If you spin two times and find the sum, what are the chances of:

3	certain	(possible)	not possible
0	certain	possible	not possible
5	certain	possible	not possible
2	certain	possible	not possible

Think and Tell!

Is any sum certain? Why?

Need: Activity Master 5, 2 paper clips for spinners.

Bears & Games
Teaching Notes

Materials Needed: Paper • Bears • Paper Clips • Red/White Counters (Chips) • Blank Dice • Red and White Dots for Dice • Activity Master 3

Warm-Up

To prepare for the first three activities, ask children to vote for their favorite TV shows and consider what information is given by:

•**the show of hands**

•**a tally based on the show of hands**

Challenge them to think about different ways of organizing and presenting the information, such as listing the shows in order from most popular to least popular; making a graph showing how many children like each show; or discussing whether one category of shows (game shows, comedies, cartoons) is more popular than others.

The objective of the last four activities is to explore games that are fair and not fair and provide opportunities for children to reflect on these concepts.

Introduce the last four activities on games by showing the class a die with 4 red dots and 2 white dots. Break the class into two groups. A group gets one point every time the group's color comes up. The first group to get five points wins the game. On the first round, let Group 1 choose the color (red or white). On the second round, Group 2 chooses the color. Following the games, have the groups discuss: whether the game is fair (it is not fair; red is twice as likely to win as white); and how the game can be made fair (change one red dot to white, so the die has 3 red dots and 3 white dots).

Using the Pages

Favorite "Bear" and **Favorite "Bear" Bar Graph** *(pages 42 & 43)*: In these two activities, children use tally marks and bar graphs to present data.

What Would the Bear Be? *(page 44)*: Children use their own data from the two previous activities to make a "best guess" or most likely pick from additional or missing data.

Is It Fair? and **Which Die?** *(pages 45 & 46)*: In these two activities, children have opportunities to play games using an unfair die.

Make It Fair and **Pick a Spinner** *(pages 47 & 48)*: Children make fair spinners and point out differences between fair and unfair spinners.

Wrap-Up

Following the activity "Pick a Spinner," urge children to make a fair die for The Bear Game (page 45).

Favorite "Bear"

Name_____

 Explore   Which bear does the class like?

Panda

Polar

Black

Koala

Favorite "Bear" Activity

Mark your tally as you are called.

Last one up: write the totals.

Favorite "Bear" Tally

Panda	Polar	Black	Koala
IIII II	III	I	IIII
Total			

Talk About It!

How did tallying help answer the question?

Think and Tell!

How would you collect this data for the whole school?

Need: 1 sheet of paper for each child.

Data, Chance, and Probability Activity Book, 1-3
© 1992 Learning Resources, Inc.

Favorite "Bear" Bar Graph

Name_____

 **Explore** Which bear did you pick?

Draw and color a picture of the bear.

Class Bar Graph

Take your Favorite "Bear" picture.

Stand in line for your bear to make a people graph.

Stand on your picture.

Now sit down.

Look at the bar graph.

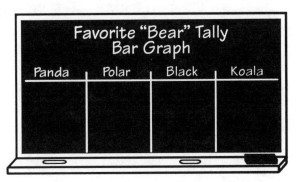

Favorite "Bear" Tally Bar Graph

Panda | Polar | Black | Koala

What does the bar graph tell you?

 Think and Tell!

How is the bar graph like the people graph and the tally?

Need: 1 sheet of paper for each child, Favorite "Bear" Tally (page 42).

What Would the Bear Be?

Name_____

 **xplore** If someone didn't vote on the bears, which is he or she:

Most likely to pick? (circle one)

panda   polar   black   koala

Least likely to pick? (circle one)

panda   polar   black   koala

Why do you think so?

How does the graph in the Favorite "Bear" Activity help you decide?

Need: Favorite "Bear" Bar Graph.

Data, Chance, and Probability Activity Book, 1-3
© 1992 Learning Resources, Inc.

Is It Fair?

Name_____

xplore   ## The Bear Game

Pick who will spin.
Spin.
If shaded, take a bear.
If not, partner takes a bear.
Repeat 9 more times.
Who will win?

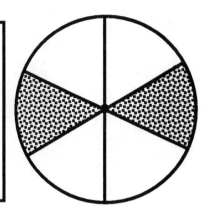

Talk About It!

Did the game turn out as you thought?

Think and Tell!

Why do you think so?

Need: Supply of bears, paper clip for spinner.

Data, Chance, and Probability Activity Book, 1-3
© 1992 Learning Resources, Inc.

Which Die?

Name_____

 **xplore**   Pick a red or white chip.

Put the chip on the start box.

Which die is best for you?

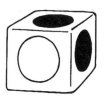 

5 red dots   **5 white dots**
1 white dot   **1 red dot**

Race Home Game

Pick red or white. Put chips on start.
Take turns. Roll the die.

If red, move one space.

If white, move one space.

Play Again

Play again using second die.

Did the games turn out as you thought?

Which die is best for the red chip? White chip?

> Need: 2 red/white chips; 2 blank dice; 6
> red, 6 white dots; Activity Master 3.

Make It Fair

Name_____

**xplore** Color the spinner to make it fair.

Play *The Bear Game.*

The Bear Game

Pick who will spin.
Spin.
If shaded, take a bear.
If not, partner takes a bear.
Repeat 9 more times.
Who will win?

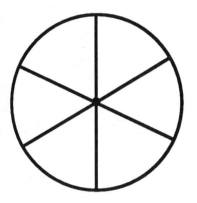

Talk About It!

Is the game fair?

Think and Tell!

Why do you think so?

Need: Supply of bears, paper clip for spinner.

Data, Chance, and Probability Activity Book, 1-3
© 1992 Learning Resources, Inc.

Pick a Spinner

Name _____

 **xplore**   Pick a spinner for *The Bear Game.*

The Bear Game

Pick who will spin.
Spin.
If shaded, take a bear.
If not, partner takes a bear.
Repeat 9 more times.
Who will win?

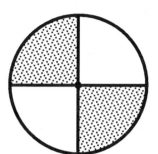

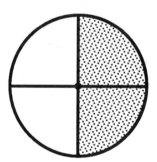

Talk About It!

Is your spinner fair?

Why do you think so?

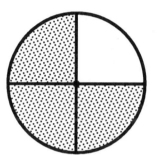

Think and Tell!

How many fair spinners are shown?

Tell how you know.

Need: Supply of bears, paper clip for spinner.

Gumballs & Gloves
Teaching Notes

Materials Needed: Red/White Counters (Chips) • Small Paper Cups • Activity Master 2

Getting Ready

This section involves a number of activities associated with a gumball machine. If possible, bring in a gumball machine and discuss how it works.

Chips are used to model gumballs or gloves in all of the activities. Because gumballs are not replaced between picks, a result is that as gumballs are drawn, the probability of a certain color on the next draw may change. Encourage the children to notice this important result for the anticipated likelihood of specific colors. When only one color is left, the probability of that color occurring on the next draw is certain.

Using the Pages

White Comes Out *(page 50)*: When children work with the red/white chips, tell them that the white needs to come out first. Then red will be certain on the next draw.

Two Come Out *(page 51)*: As in the "White Comes Out" activity, students need to know that the colors for the first two chips are predetermined, that is, white follows red. This means that red is drawn followed by white in the gumball machine.

Pick Two *(page 52)*: The three possible patterns for the gumball machine are RRW, RWR or WRR (R is red, W is white). Urge students to discuss these patterns. Some children may observe that red is more likely on the last draw.

Two Gumball Jars *(page 53)*: This exercise extends the concept of patterns. Jar number two has four different color patterns because the gumballs are the same color.

Lost and Found *(page 54)*: Help children discover that three draws are needed to ensure getting a pair. Repeat this exercise until children understand.

Which Gumball Jar? *(page 55)*: Make a simple gumball machine with a paper cup. Cut a 1-inch slit along the cup rim. Place the chips (2 red and 1 white up) on a table under the cup. Rotate the cup several times *without lifting it off the table* until a chip comes out.

Which Jar? *(page 56)*: As an extension activity, tell children to make up their own "best jar" problems. Ask each child to give his or her problem to a partner to solve.

Wrap-Up

You might encourage the children to make other models (e.g., die, spinner) for the gumball machine or even make their own type of gumball machine.

White Comes Out

Name_____

 **xplore**   The machine has 3 gumballs.

A white gumball comes out.

What color will come
out next time?

Your Turn

Play the gumball machine with chips.

Refill the gumball machine so your partner can play.

Did the games turn out as you said?

Was the color red certain to come out?

Why do you think so?

Need: 3 red/white chips, Activity Master 2.

Two Come Out

Name_____

 **xplore**   The machine has 3 gumballs.

Red comes out first.

A white gumball comes out next.

What color will come
out next time?

Your Turn

Play the gumball machine with chips.

Refill the gumball machine so your partner can play.

Did the games turn out as you said?

Was the color red certain to come out?

Why do you think so?

Need: 3 red/white chips, Activity
Master 2.

Data, Chance, and Probability Activity Book, 1-3
© 1992 Learning Resources, Inc.

Pick Two

Name _____

 **xplore**   The machine has 3 gumballs.

Two gumballs come out.

What color could come out next time?

Your Turn

Play the gumball machine with chips.

Refill the gumball machine so your partner can play.

Can the colors come out any other way?

Was any color certain for the third chip?

Why?

Need: 3 red/white chips, Activity Master 2.

Two Gumball Jars

Name_____

 **Explore** Get 4 gumballs from jar 1.

What colors did you get?

Jar 1

Get 4 gumballs from jar 2.

What colors did you get?

Jar 2

Your Turn

Play the gumball machine with chips.

Refill the gumball machine so your partner can play.

Did it turn out as you thought?

How many patterns are there for Jar 1? Jar 2?

Need: 8 red/white chips, Activity Master 2.

Data, Chance, and Probability Activity Book, 1-3
© 1992 Learning Resources, Inc.

Lost and Found

Name_____

 **xplore**   Don't peek.

Take 1 glove, then another...

How many gloves did you pick before you found a pair?

Your Turn

Play lost and found with chips.

Replace the chips so your partner can play.

How many chips did you pick before you were certain to get a pair?

Why?

Need: 4 red/white chips (use for gloves).

Data, Chance, and Probability Activity Book, 1-3
© 1992 Learning Resources, Inc.

Which Gumball Jar?

 **xplore** You have one chance to get a red gumball. Which jar is best for you?

Your Turn

Choose a jar.

Play the gumball machine with chips.

Refill the gumball machine so your partner can play.

 Did the best gumball machine turn out as you said?

 Why is one jar better than the other for picking a red chip?

Need: 6 red/white chips; paper cup for gumball machine.

Which Jar?

 **Name**_____

 **xplore** You have one chance to get a red gumball.
Which jar is best for you?

Your Turn

Choose a jar.

Play the gumball machine with chips.

Refill the gumball machine so your partner can play.

 Did the best gumball machine turn out as you said?

 What would be even better?

Color the gumballs.

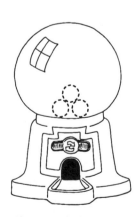

Need: 6 red/white chips; paper cup
for gumball machine.

Data, Chance, and Probability Activity Book, 1-3
© 1992 Learning Resources, Inc.

Spinners & Sums
Teaching Notes

Materials Needed: Red/White Counters (Chips) • Paper Clips • Red and Yellow Blocks or Paper Squares • Masking Tape • Activity Master 3

Warm-Up

Many activities in this section use two spinners to introduce children to *combining outcomes*. As a lead-off, have students take turns rolling two regular dice and recording the sum. Have them continue until they are convinced that all possible outcomes from 2 to 12 have been listed. After examining "best chance" and "fair chance" situations, the activities of this section focus on combining spinner outcomes.

Using the Pages

Which Spinner? *(page 58)*: Data produced by each pair should confirm that spinner 1 is better for "red" and spinner 2 is good for "white." If not, pool the data from all pairs.

Which Has a Better Chance? *(page 59)*: It is possible that the most likely color will not turn up most often. However, if class results are pooled, the expected prediction should occur. Such situations provide opportunities for children to recognize that unlikely events do occur.

What is Best? *(page 60)*: Children may not need to play the games before writing about them.

Spin and Graph *(page 61)*: Urge children to think about the chances of the two colors. Have them use the words "twice as likely," "two times more likely," "two to one," and "twice as often." Note: Some students may see that the odds in favor of the shaded color are two to one.

Fill In *(page 62)*: In the "Think and Tell" section, children should discover that the "O" in the box will ensure a *certain* event.

What About the Sum? *(page 63)*: If a child picks 1 for the box, the sum of 3 is certain. Clearly, this also means that 3 does have a chance — in fact, a 100 percent chance.

Pick a Sum *(page 64)*: Review "What About the Sum?"(page 63) before starting this activity.

Wrap-Up

Have children reconsider the total set of sums that occurred when they rolled two regular dice in the "Warm-Up" to this section. Questions like the following should be considered: 1) What sum has no chance? (1 has no chance.) 2) Is any sum *certain*? (No.) 3) Which sum is most likely? (Sums of 7 account for 6 out of the 36 possibilities. Seven is the most likely sum because the greatest number of combinations are possible, namely, 1 and 6, 6 and 1; 2 and 5, 5 and 2; 3 and 4, 4 and 3.)

Which Spinner?

Name_____

 xplore   Each player picks one chip, either red or white.

Put your chip in the start box for the *Race Home Game.*

Which spinner is best for you?

Race Home Game

Pick red or white. Put chips on start.
Take turns. Spin.

If red, move one space.

If white, move one space.

Play again. Use spinner 2.

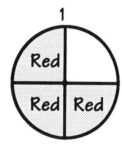   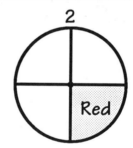

Did the game turn out as you said?

Which spinner is better for the:

red chip?   white chip?

Need: 2 red/white chips, paper clip for spinner, Activity Master 3.

Which Has a Better Chance?

Name_____

 **xplore** Think about playing the *Race Home Game* again.

Which game chip has a better chance? Circle one.

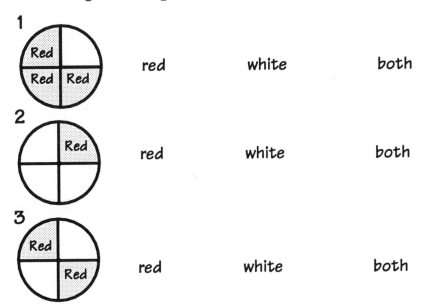

1   red   white   both

2   red   white   both

3   red   white   both

 **Talk About It!**

How did you decide which color has a better chance of winning?

 **Think and Tell!**

Are spinners 1 and 2 fair or not fair?

Why do you think so?

Why aren't spinners 1 and 2 used in real games?

Need: 2 red/white chips, paper clip for spinner, Activity Master 3.

What is Best?

Name_____

 **E**xplore In the *Race Home Game* you have a white chip in the start box.

Game 1   Game 2   Game 3

 How would each game turn out using your white chip?

Game 1 _____

Game 2 _____

Game 3 _____

 Which spinners are fair?

Which spinners aren't fair?

Why do you think so?

> Need: Activity Master 3.

Spin and Graph

Name_____

 **xplore**   Take turns.

Spin the spinner.

Take a block to match the color.

Spin and repeat 9 times.

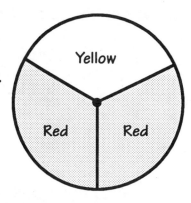

Class Activity

Make a red-yellow class bar graph to show your results.

Then make a tape strip as high as the yellow stack.

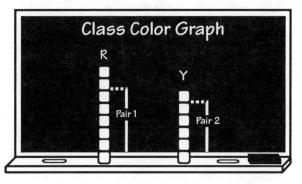

About how many yellow tape strips are needed to match the height of the red?

How are the spinner and the bar graph alike?

What spinner color has a better chance?

Need: 18 yellow and 18 red blocks or paper squares; paper clips for spinner, masking tape.

Fill In

Name_____

 **E**xplore   Pick a number: 0, 1, or 2.

Write it in the ☐ .

Spin each spinner to get two numbers and find their sum.

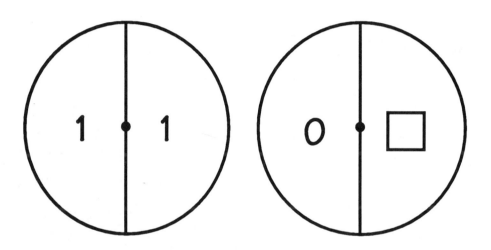

 Circle what sums are:

certain   possible   impossible

Think and Tell!

What number in the ☐ would give a certain outcome?

Need: Paper clip for spinner.

Data, Chance, and Probability Activity Book, 1-3
© 1992 Learning Resources, Inc.

What About the Sum?

Name _____

 **xplore** Write numbers in the spinners.

Spin to get two numbers and find the sum.

Try it to check.

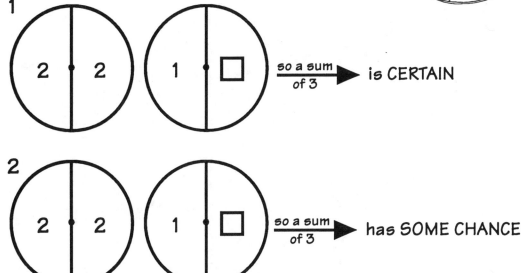

1   (2 | 2)   (1 | □)   so a sum of 3 → is CERTAIN

2   (2 | 2)   (1 | □)   so a sum of 3 → has SOME CHANCE

 **Talk About It!**

How did you choose the numbers for the boxes?

 **Think and Tell!**

Use the two spinners.

What sums are possible?

What sums are most likely?

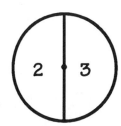   (2 | 3)   (1 | 2)

Need: Paper clip for spinners.

Data, Chance, and Probability Activity Book, 1-3
© 1992 Learning Resources, Inc.

Pick a Sum

Name_____

 **xplore** Write numbers in the spinners.

Spin to get two numbers and find the sum.

Try it to check.

1

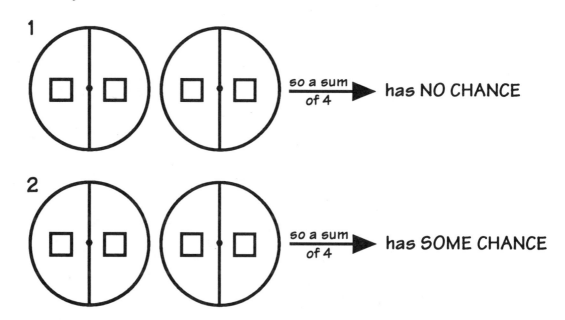

so a sum of 4 ▶ has NO CHANCE

2

so a sum of 4 ▶ has SOME CHANCE

Talk About It!

How did you choose the numbers for the boxes?

Think and Tell!

What number in the ☐ would give a certain outcome?

Need: Paper clip for spinners.

Data, Chance, and Probability Activity Book, 1-3
© 1992 Learning Resources, Inc.

Arms Up
Teaching Notes

Materials Needed: 2" x 2" Squares • Masking Tape • Red/White Counters (Chips) • Activity Master 6

Warm-Up

Demonstrate how your arm size can be measured from the point of the elbow to the tip of the longest finger. To measure, use string or centimeter measuring tape.

Using the Pages

What's Your Arm Size? *(page 66)*: Find the median size by lining up children in ascending order of arm size and selecting the middle child. In the case of an even number, you can join in the activity so the total number will be odd.

Double Bar Graph *(page 67)*: In comparing groups, tell children to look back at the data involved in the Word Box in the "What's Your Arm Size?" activity.

What Would the Arm Size Be? *(page 68)*: The most frequent arm size of the girls is the "mode" for the girls from the activity "What's Your Arm Size?" Similarly, the "mode" for the boys is the most frequent arm size of the boys. Children might compare the two modes and explain how they affect their predictions for a boy or a girl.

Arm Fling *(page 69)*: Assist children who need help placing their target sheets. Use masking tape to make the tossing line.

Who's in the Middle? *(page 70)*: When finding a median, join in the activity if needed to make sure the total number is odd. To post scores, children can write on the chalkboard (in order, low to high) or tape papers to the board or wall. Note there is only one middle score, but several children may obtain that score. Also, there may be more than one "most frequent" score.

Wrap-Up

Challenge students to come up with ideas similar to the "Arms Up" activity. These might be characteristics typical of children of the same age. Data can then be used to compare several age groups, for example, first graders and fourth graders, first graders and third graders.

What's Your Arm Size?

Name_____

 **Explore** Measure each other's arm — elbow to finger tip — to the nearest centimeter.

Record on your paper squares.

What's Your Arm Size?

Girls and boys work in separate groups.

Each lines up in order of arm size.

Those with the same size arms double or triple up.

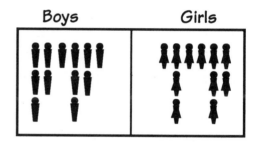

| Boys | Girls |

Talk about each graph.

Word Box

Range (smallest-biggest)

Middle score (median)

Most frequent (mode)

How do the two groups compare?

Need: One 2" x 2" paper square for each child (different color for boys and girls).

Data, Chance, and Probability Activity Book, 1-3
© 1992 Learning Resources, Inc.

Double Bar Graph

Name_____

 xplore Girls and boys work in separate groups.

Each finds the smallest and biggest arm sizes.

Graph Activity

Use your squares to make a double bar graph for boys and girls.

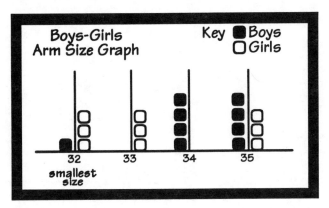

Boys-Girls Arm Size Graph

Key ■ Boys □ Girls

32 — smallest size   33   34   35

 **Talk About It!**

Does the double bar graph help you to compare boys with girls?

 **Think and Tell!**

How do the two groups compare?

Need: Paper squares from *What's Your Arm Size?* (page 66); tape to attach squares to board.

What Would the Arm Size Be?

Name_____

 **xplore** If a girl missed the arm size activity, what is her arm size most likely to be?

About ___cm.

 Talk About It!

Would it be the same for a boy?

Why do you think so?

 Think and Tell!

What helped you to decide?

Need: Double bar graph and Word Box data from pages 66-67.

Arm Fling

Name_____

 **xplore**   Set up the Arm Fling Target.

Stand 3 paper lengths away.

Toss the chip 3 times.

Record your total score.

Data Activity

What total scores are possible?

Work with your partner to list them, low to high.

Use the Word Box to write about your list.

```
┌─────────────────────────────────┐
│          Word Box               │
├─────────────────────────────────┤
│   Range (smallest-biggest)      │
│     Middle score (median)       │
│     Most frequent (mode)        │
└─────────────────────────────────┘
```

Was your list the same as others in the class?

Need: Red/White chips, Activity Master 6.

Who's in the Middle?

Name_____

Explore What was your total score in the Arm Fling?

Write it in the ☐ .

Class Activity

Line up, low to high, by total score.

Post your score.

| 0 | ☐ | ☐ | - - - - | ☐ | ☐ | 15 |

Talk About It!

Use the word box to tell about the class scores.

Word Box

Range (smallest-biggest)

Middle score (median)

Most frequent (mode)

Think and Tell!

Is there only one middle score?

Did one score come up more often?

Need: Paper square for each child.

Activity Master 1
Heads or Tails

Activity Master 2
Gumball Machine

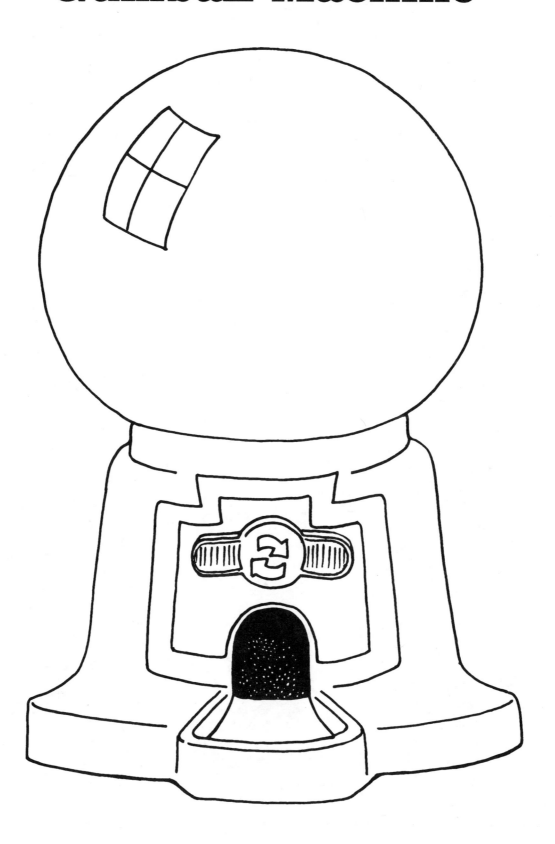

Data, Chance, and Probability Activity Book, 1-3
© 1992 Learning Resources, Inc.

Activity Master 3
Race Home Game

Activity Master 4
Match It Cards

Data, Chance, and Probability Activity Book, 1-3
© 1992 Learning Resources, Inc.

Activity Master 5
Spinner Cards

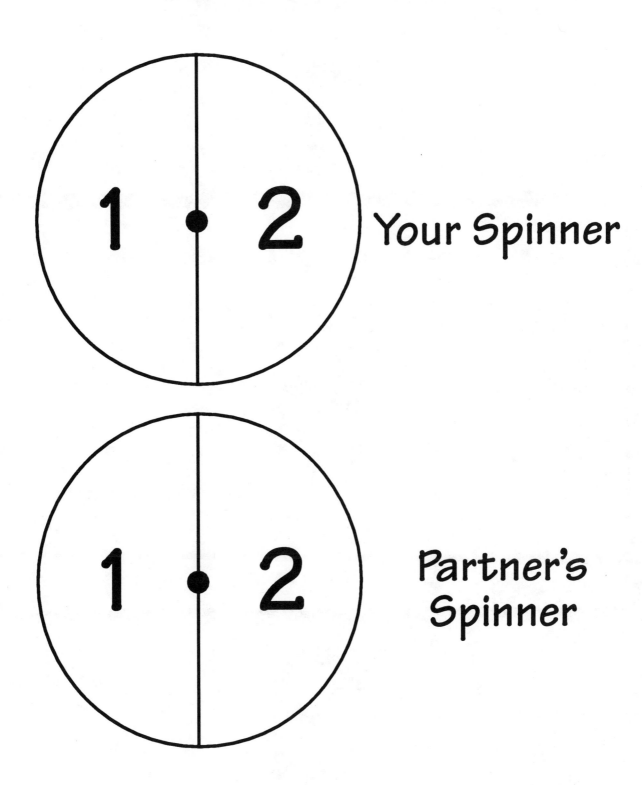

Your Spinner

Partner's
Spinner

Activity Master 6
Arm Fling Target

1

5

2

Data, Chance, and Probability Activity Book, 1-3
© 1992 Learning Resources, Inc.

Activity Master 7
Bears

Progress Chart

Name_____   **Date Started**_____

Grade_____   **Date Finished**_____

Bears & Spinners

- ☐ Bears in a Bag
- ☐ Pick a Bear
- ☐ Pick Again
- ☐ Pick One More Time
- ☐ Spin and See
- ☐ Spin and Tell

Hide & Spin

- ☐ Which Color?
- ☐ Which Spinner Color?
- ☐ Which Color Now?
- ☐ Which Spinner?

Bears, Books, Crayons, & Coins

- ☐ Story Time
- ☐ What Would the Story Be?
- ☐ In the Library
- ☐ What's in the Bag?
- ☐ Best Chance
- ☐ Second Best
- ☐ Which Spinner for the Bears?
- ☐ What Will It Be?
- ☐ Heads or Tails?
- ☐ Match the Coin

Game Time

- ☐ Is It Fair?
- ☐ Make It Fair
- ☐ Which Spinner Color?
- ☐ Red or White?
- ☐ Try Again!
- ☐ Get a Match!
- ☐ Spin and Sum

Bears & Games

- ☐ Favorite "Bear"
- ☐ Favorite "Bear" Bar Graph
- ☐ What Would the Bear Be?
- ☐ Is It Fair?
- ☐ Which Die?
- ☐ Make It Fair
- ☐ Pick a Spinner

Gumballs & Gloves

- ☐ White Comes Out
- ☐ Two Come Out
- ☐ Pick Two
- ☐ Two Gumball Jars
- ☐ Lost and Found
- ☐ Which Gumball Jar?
- ☐ Which Jar?

Spinners & Sums

- ☐ Which Spinner?
- ☐ Which Has a Better Chance?
- ☐ What is Best?
- ☐ Spin and Graph
- ☐ Fill In
- ☐ What About the Sum?
- ☐ Pick a Sum

Arms Up

- ☐ What's Your Arm Size?
- ☐ Double Bar Graph
- ☐ What Would the Arm Size Be?
- ☐ Arm Fling
- ☐ Who's in the Middle?

78

Data, Chance, and Probability Activity Book, 1-3
© 1992 Learning Resources, Inc.

Data, Chance & Probability

GOOD WORK AWARD

To: _____

For: _____

Date

Teacher

14 2 10 3 18 6 13
16
12
15
0
17 3 8 10 7 1 20
11 6 12 2 19 4 5 0
9

FAMILY-GRAM

Dear Family,

Congratulations! Your child has successfully completed the unit entitled

_____

in Data, Chance, and Probability Activity Book, 1-3

_____   _____
Date   Teacher

Data, Chance, & Probability
Award Certificate

Congratulations to: _____

for excellent work in the unit entitled

_____

in Data, Chance, and Probability Activity Book, 1-3

_____   _____
Date   Teacher